The Human Operating System

The Key to the human development and self-growth science

Book 1

Mohamed Amer

While every precaution has been taken in the preparation of this book, the publisher assumes no responsibility for errors or omissions, or for damages resulting from the use of the information contained herein.

HUMAN OPERATING SYSTEM

First edition. July 27, 2018.

Copyright © 2018 Mohamed Amer.

Written by Mohamed Amer.

Table of Contents

Introduction ____ *1*

First: The Biological Operating System ____ *4*

    1. The system of Nutrition and Energy ____ 5

    2. The system of Movement, Protection and Support ____ 7

    3. System of Control ____ 8

    4. Filtration, Purification, and Waste Disposal System ____ 9

    5. The system of Regeneration, Reproduction and Conservation of Species ____ 10

Second: Psychological Operating System ____ *11*

    1. The system of Nutrition and Energy ____ 14

    2. The system of Movement, Protection and Support ____ 16

    3. System of Control ____ 17

    4. Filtration, Purification, and Waste Disposal System ____ 18

    5. The system of Regeneration, Reproduction and Conservation of Species ____ 19

Conclusion and introduction for the coming book ____ *21*

Dedication

To my teacher, father and mother ...
To sisters, brothers and my wife ...
To my sons and my grandsons ...
To all people all over the world ...
 I dedicate this book to you all.

Introduction

Man is the greatest creature on the face of the earth. He is the intellectual being who can subject all other beings to serve him. The greatest thing in this man is his mind by which he reached the moon, dove into the depths of the oceans, and achieved the scientific miracles.

Today, we enjoy these achievements as man managed to communicate with the whole world instantly in voice and image. Despite these numerous and great achievements, wars, tragedies, suffering, pains, and diseases are still increasing in all corners of the world; we have not been able to put an end to them yet, let alone the fact that we most of the times are the cause of them.

If man, who made the computer in a stimulation of the human mind, has reached the latest operating systems of this computer, so what is the most efficient human operating system that can repair man and be suitable for him?

And also in your own selves; can you not see? (Quran 51:21)

Man consists of spirit, body, and soul. The soul is the product of spirit's residency in the cage of the body. Soul, therefore, tastes death when the spirit exits this cage. Spirit is such a bird that if it lives in the cage of the body, it gives it value and gifts it with life. For this reason, man spends most of his life serving this cage in which the spirit lives, and when the spirit leaves it, the cage loses its value and the source of its existence and burial becomes the only way to honor such a cage. There is no value to the largest factory in case it lost its source of power and the capacity by which it operates. Similarly, this is the case with the largest lamp or lighting device when the power outages occur, or the most powerful communication device when the signal is cut.

The spirit for the body is just like the signal of communication or the electric current that provides the devices and equipment with energy and power. The state of death is similar to that of the machine after being unplugged or the communication device when it becomes out of coverage.

Whereas man's life is limited, and he has an inevitable end no matter how long he lives, and whereas every soul will taste death, why does not man, from the cradle to the grave, seek the best human operating system by which he can invest in his life very well and gain the most useful benefits for him, his family, community, homeland, nation and the world as a whole?

If man's soul is his inner self that includes: knowledge and ignorance, light and darkness, righteousness and debauchery, good and evil, truth and falsehood, war and peace, what would be the best investment and in what fields would it be better and more beneficial?

If all wise and sane people agree that the investment in the fields of knowledge, light, righteousness, goodness, truth and peace are the best and most feasible, how then can we invest the best investment in these fields while the conflict between these contradicting powers exists and is going on even in man's inner soul? This will not happen unless we succeed in seeking and identifying the optimal human operating system that resolves the conflict in man's soul to begin with so as to enable such a man to resolve it afterward in his family, society, homeland, his nation and the world as a whole.

First: The Biological Operating System

In ancient times, the human body was among the strange and mysterious things that scientists could not explain. However, with the scientific and technological development, and the development of medical devices, scientists were able to know the compositions of the human body, which facilitated the identification, control and treatment of many diseases.

The human body is so complex and works in a balance among its parts without malfunction. The human body consists of the various cells that form the basic building units in the body. The cells of the same type are connected together to form the tissues, while the tissues of different types are connected to form the organs which, in the process, are connected together to form the apparatuses of the human body. The human body consists of a group of apparatuses each of which is specialized in certain functions in the body. These systems are interconnected and function in a harmonious and homogenous way to achieve the complete functional integration among all organs and other systems in the body. This services the interests and welfare of this living creature.

The human body needs water, food, and oxygen as it needs protection and communication with the environment as basic conditions of life. All human body apparatuses are integrated together to achieve these conditions. We call this the biological operating system which is categorized according to the following systems:

1. The system of Nutrition and Energy

THE DIGESTIVE SYSTEM function is to take water and nutrients necessary for the body from food. Man eats the food in its complex form and cannot benefit from it directly, so he digests the food analyzing it into simple particles that can be absorbed by the cells and then discharges the useless parts to get rid of them. The digestive apparatus is integrated with the respiratory apparatus responsible for obtaining oxygen necessary to sustain the life of the cells in the body. When oxygen enters the body and interacts with the food, it produces the energy necessary to the body and consequently, the gases that body does not need, such as water vapor and carbon dioxide, goes out. The respiration process in the human body is the use of oxygen necessary to produce energy from food. The endocrine system, which is one of the most important apparatuses that controls the various functions in the body such as growth, movement and reproduction, contributes, as well, to the nutrition process in addition to its role in regulating the metabolism processes responsible for the production of energy within the body cells. This happens through catabolism of the nutrients digested by the digestive apparatus and converting them to different forms of energy through a chain of chemical reactions.

Metabolism is the process through which different cells and tissues are built and destroyed. The main function of which is to provide the necessary energy or calories to cover the functions of the body or to maintain the vital functions, so it serves as a regulator and director of energy in the body. As for the circulatory apparatus, responsible for transferring the blood to all cells, serves as a link between all apparatus because it carries food and oxygen to the cells, transfers the carbon dioxide and waste from the cells to be discarded, and transfers the hormones produced by the endocrine system. The circulatory apparatus is operated by the heart, which is a fist-sized muscle that functions as a pump of blood.

2. The system of Movement, Protection and Support

THE SKELETAL APPARATUS is responsible for giving the stiffness and the external structure of the body through the bone structure which includes bones and joints in the body.

The bones constitute a complex system consisting of many cells, fibers, proteins and minerals. The bone structure serves as the scaffold of the outer bodyguard for the body by providing support and protection for the soft tissues that form the rest of the body. New blood cells are produced from the red marrow inside our bones. The real problem that our bone structure may face, as the body's bodyguard, is that it may be broken. What mitigates this problem is the ease with which most of the bone fractures can be treated. It is just a matter of time until the healing occurs, unlike other apparatuses which their damage can lead to man's death.

As for the muscular apparatus, it is responsible for processing the movement through muscles, tendons and ligaments according to an integrated system all over the body. It is the source of movement invertebrates in general with the help of both the nervous system and the skeletal system. The nervous system regulates the movement, but some muscles such as the heart are self-regulating. Plus, the muscles are based on the skeletal system, and when there is a movement, there are at least two muscles performing this movement: one of them is constricted, i.e., shortened and becomes thicker in the middle, the other is relaxed, i.e., grows longer.

All apparatuses are covered and enwrapped by the integumentary system of exocrine glands, which consists of the skin glands, hair, and nails. Skin serves as the outer cover of the body and forms a barrier to protect the body from chemical substances, diseases, ultraviolet radiation and physical damage.

Protection and internal support are addressed by the immune system and lymphatic system by washing the cells of the body tissues and supplying them with food through clear watery fluid called lymph which separates from blood and then returns to it. The function of this apparatus is to produce the white blood cells responsible for protecting the body against diseases. The immune apparatus and the lymphatic system in the body are closely related. If we look closely, we will see that they share the physiological functions of several apparatuses. The immune apparatus is the body's defense system against viruses that cause infectious diseases, bacteria, fungus and parasitic animals. The immune system keeps these harmful factors outside the body and attacks those that are able to enter therein.

3. System of Control

THE NERVOUS APPARATUS is the controller and the giver of orders to the various apparatuses of the body as well as communicating the information among them. It is the organizer inside the body. The nervous system consists of the brain, the spinal cord that affects the senses. All nerves connect these apparatuses with the rest of the body. It is responsible for controlling the body and for communicating and connecting its parts. Brain and spinal cord form the control center knowing as the central nervous system (CNS) which assesses the information and the decisions taken.

4. Filtration, Purification, and Waste Disposal System

THE FILTRATION, PURIFICATION, and waste disposal process are performed through the excretory system responsible for the disposal of harmful compounds and materials in the body resulting from the continuous process of metabolism. Some of the materials are disposed through the porosity of the skin by sweat glands under the surface of the skin, through lungs during exhalation process, through liver and bile and some other materials are discharged through the intestines in form of fecal wastes.

As for the urinary apparatus, it is the main excretory system and is primarily responsible for the process of discharging waste and urotoxia out of the body. The urinary system consists of kidneys, ureters, urinary bladder, and urethra. The function of the kidney is to purify the blood by removing wastes and producing urine. The ureters, urinary bladder, and urethra together form what is known as urinary tract, which acts as a plumbing system that drains urine from kidneys, collects it, then releases it through urination process. The importance of the urinary system lies in regulating the level of blood pressure in the body, the level of water lost during urination, and improving the concentration of potassium and sodium ions in the body as well as maintaining the hydrogen number in the blood, and discharging the excess urotoxia and waste in the body and removing them out.

5. The system of Regeneration, Reproduction and Conservation of Species

THIS HAPPENS THROUGH the reproductive apparatus responsible for human reproduction process which ensures the conservation of human species through sexual reproduction. The reproductive apparatus of female and that of the male are the ones responsible for the production of embryos.

Second: Psychological Operating System

Psychology is the science of studying the general behavior of the living creatures in all external forms of movement and the internal forms of mentality and psychology. The most important pivots that this science investigates and researches are all behaviors that emanate from the human being and the internal and mental activities such as the process of thinking, as well as the emotional side and emotional effects such as feelings, sadness, joy, happiness, fear, etc. Psychology is considered one of the ancient modern sciences. It first appeared as a branch of the philosophical sciences. In the past, man sought to understand the human phenomena and the human soul. To meet this human need, philosophers sought to develop some perceptions of human existence, knowledge and mind. Many of psychological philosophical theories emerged from these perceptions. Such as:

the theory of knowledge, the theory of existence, values, philosophy of reason and others.

Recently, with the expansion of knowledge in various sciences in general and in psychology in particular, a dire need has emerged for psychology to be full-fledged cognitive and independent science. Like humanities, psychology, in general, has three objectives: understanding, prediction and control because psychology seeks to control the behavior through understanding the behavior and predicting it from the available data. This occurs by identifying the stimuli and their associations with the various behavioral responses, i.e., Predicting the time of behavior and controlling some of the independent variables that cause a particular phenomenon. Thus, psychology predicts the time of behavior and when it may happen, i.e., the assumption of certain behaviors in case some stimuli took place which consequently leads to the emergence of expected behavioral responses.

Psychology tackles a single subject: the study of the psychology of human beings and their internal secrets; the attempt to explain their behavior and reaction in different situations; forming a complete image of the psychological state of the individual and discovering its general characteristics and qualities. Psychology is not the study of the soul or one's self as may be understood from the name, but it studies the human being as a whole because the human soul is the man himself in all components. The mental health is defined as the relatively permanent state of the individual in all mental, psychological and physical aspects, free from organic diseases and mental disorders, so the individual enjoys a state of psychological, personal and social harmony followed by a sense of comfort which enables man to prove himself and exploit the opportunities available to achieve his goals.

Mental illness is opposite to mental health and it can be defined as a psychological disorder that results in unfavorable attitudes of the individual toward himself and his society, i.e., a state of disharmony in psychosocial interaction and emergence of abnormal behaviors that affect his achievement and affect the safe and stable life of himself and of those around him. The mental illness in some advanced psychiatric conditions may lead to organic or functional disease, such as psychological respiratory disorders or neurodermatitis

Undoubtedly, there is a precise and close relationship between the terms we coined: the biological operating system and the psychological operating system. As long as man is a combination of spirit and body and man's soul is the product of the spirit that lives in the cage of the body, there must be a conflict between the needs of the spirit and the needs of the body. as the human being needs water, food, oxygen, protection, and communication with the surrounding environment as basic conditions for his life, all apparatuses of the body are integrated together to fulfill these conditions, and this what we defined as the biological operating system. In the same process, in addition to the food that body needs, that bird which lives in the cage of the body, i.e., the spirit needs food as well. Life of human being will not stabilize and be balanced unless there is a balance and integration between the food of the spirit and the food of the body to ultimately serve man's soul and this what we called the psychological operation system. For this reason, we will adopt the same categories of systems and apparatuses used the biological operating system in the psychological operating system.

1. The system of Nutrition and Energy

WHEN A HUMAN BEING is a helpless fetus in his mother's womb, he is fed in the womb from the food and energy of his mother. Once he is a newborn baby, he needs food as well as protection, care, mercy, love and compassion as he begins the quest to know himself and the surrounding environment. While the digestive system obtains food and nutrients necessary for the body from the food, man's soul spontaneously beings to seek the knowledge necessary to know himself as an independent living creature that has psychological and physical needs and requirements, as well as talents, abilities and faculties. He also seeks to explore the external environment from which he acquires most of his knowledge and experiments that are gradually digested in accordance with the stages of growth. And when he grows up to be in complete physical structure, and starts to feel independent and free, he begins to contemplate and reflect on everything, even on the mystery of his existence and his role, duties, and mission as a human being experiences the different conditions in the struggle to survive and to exist in this life.

You, as a rational human being with full power of choice, cannot, while walking daily in the many and varied paths of life, take the simplest decision such as saying: I will eat such or such food, purchase that thing or the other thing, go to such a place to do such and such, do this thing for such a purpose, without having the primary or minimum knowledge of these things. This knowledge is the theoretical science that you should seek in all walks and aspects of life. The more this knowledge expands in yourself, the more your experiences expand in this life. However, all these choices are theoretical to you even if they are regarded as applied sciences and practical knowledge done by others. Practical experience will not be gained except from the choices you passed through and practically experienced by you. Thus, you try them yourself and then you are able to classify them as beautiful or ugly, good or bad, and beneficial to mankind and its usefulness remains in the earth or useless like the foam of the sea that vanishes away. The more you combine the theoretical and practical knowledge, the more you gain knowledge by which you are enlightened, your behavior is evaluated and assessed, and your decision is corrected. All of these accumulated sciences, experiences and knowledge are things that provide you with energy and ability; contribute to forming your personality and serve as a scientific reference, and the logical justifications and reasons for your decisions, and by which you make the right choice when tested and when you take a decision. Thus, they are saved gradually until the time comes to take any decision through classifying your theoretical sciences by comparing them to your personal and practical experiences, and by benefiting from those whom you trust of family, friends, wise and rational people from your age or the previous ages. Knowledge is the nourishment of the soul, and work is the digestive system by which we digest the science we learned and we then breathe the knowledge which is the fruit of both science and the psychic energy that we need and by which we walk in the paths of life.

2. The system of Movement, Protection and Support

THE PSYCHOLOGICAL STRUCTURE of man and its strength is manifested in his personality, which is formed, developed, and crystallized as he progresses in age and grows more maturely through sciences, knowledge, experiences, and skills that he acquired theoretically and practically such as various conditions and experiences he himself was exposed to on one hand, and his interaction with the external environment on the other hand. Man's life journey is so short compared to the human life throughout the ages. Man is by nature a social being. He cannot live in isolation from people. He interacts with his present society negatively and positively as he interacts with the history of his fathers and ancestors which stands as his belonging and his roots. In this way, man is not separated from the movement of his society as well as the movement of history and even geography. This movement inside man does not cease nor does it come to an end as long as his heart beats. The more man benefits from the knowledge he acquires by himself and supports his personal knowledge and experiences with those of others throughout the ages, the more he gains protection, immunity and support against the different currents of life.

3. System of Control

IF THE BRAIN IS THE main part of the nervous system inside the skull that collects and analyzes information, as well as manages and controls over most organs in the body and be considered the source of information production, the mind is the term that used mostly to express and describe all of higher brain functions that makes the human being conscious, such as: argument, concentration, intelligence, memory, analysis and emotion. Therefore, the mind is a word that expresses the ability to think and to reason which characterizes man from other creatures, but the brain is a material term refers to the content inside the human skull which includes brain, brainstem, and cerebellum. Although man is characterized by being a rational creature, his mind is incapable of knowing the unseen or even comprehending the changes in his heart, and incapable of knowing everything. Therefore, the mind and the heart share the duty in the control system because man can reason by his heart things he cannot reason by his mind.

4. Filtration, Purification, and Waste Disposal System

AS MAN AGES, HIS PERCEPTIONS grow and becomes more aware and more mature. This qualifies him to evaluate and assesses all his previous experiments and experiences, including those he gained from the environment in which he was born or from trials and experiences of others throughout the ages. The process of filtration, purification and waste disposal is performed by taking the useful and beneficial things and disposal of harmful wastes.

5. The system of Regeneration, Reproduction and Conservation of Species

TRIALS ENABLE US TO acquire new experiences and renewable knowledge. A wise man does not only benefits from his personal experience, but he also benefits from the experiences of people of his age and previous ages. If man in previous ages needed more strength of the body muscles, today's man is in a dire need for the strength of the muscles of mind so as to understand and comprehend the changes of his heart in conditions of life that became more comfortable and easier. Man, therefore, is no longer in need of great muscle effort, but because these conditions are becoming more complicated, problematic and changing, this makes it necessary for man to activate his mind so as to decipher their symbols and hieroglyphs. In this way, man can know and tackle himself during the process of knowing and identifying his talents, powers, abilities and faculties that help him face the requirement and conditions of life today.

Reproduction happens through benefiting from the knowledge, experiences and trials of people throughout the ages. Thus, a cross-fertilization of ideas takes place and this leads to generating new ideas. A wise man should not imprison himself in the circle of his ideas, but to be open to the ideas of people so his ideas are multiplied and reproduced by ideas of others and their knowledge, experiences and experiments. In this way, mankind develops and human being increases in experience, knowledge, progress and well-being that contribute to the development and preservation of human species.

Conclusion and introduction for the coming book

After what we put forth as introduction and general framework of the invented term "the human operating system" which, indeed, constitutes the pivot of every man's life in this universe because his life from cradle to grave revolves around the orbit of this operating system, whether he knows it as a specialist or not, or he is knowledgeable aware of it or not, we come back with you dear reader to provide an answer to the main question which will be and substance and essence of series of books following this introductory book:

If man, who made the computer in a stimulation of the human mind, has reached the latest operating systems of this computer, so what is the most efficient human operating system that can repair man and be suitable for him?

And also in your own selves; can you not see?
(Quran 51:21).

Born as a Muslim, my mind did not accept Islam just because I was born as a Muslim, nor was I convinced by the situation of Muslims today. I have been contemplating and reflecting since my childhood on this universe and seeking my duties, mission and the secret of my existence as a human being in an immense world that expands to the point that I feel like a feather in the wind, and narrows to the point that it became a small global village that I can control and manage when I control myself.

I have not found in this world a better operating system or more fantastic and beneficial to man than the Islamic operating system. The Islamic operating system will be the title of the next series of books in which we will deal with Islam as a thought and a life course not as a dogma or religion in a way that serves both Muslims and non-Muslims.

www.ingramcontent.com/pod-product-compliance
Lightning Source LLC
Chambersburg PA
CBHW031508210526
45463CB00003B/1140